Antonina Kuzhym

Technology incubator as an ecosystem module

Antonina Kuzhym

Technology incubator as an ecosystem module

Technology incubator as an organizational solution within the ecosystem of effective implementation of the innovation model

ScienciaScripts

Imprint

Cover image: www.ingimage.com

This book is a translation from the original published under ISBN 978-620-7-65437-6.

Publisher:
Sciencia Scripts
is a trademark of
Dodo Books Indian Ocean Ltd. and OmniScriptum S.R.L publishing group

120 High Road, East Finchley, London, N2 9ED, United Kingdom
Str. Armeneasca 28/1, office 1, Chisinau MD-2012, Republic of Moldova, Europe
Managing Directors: Ieva Konstantinova, Victoria Ursu
info@omniscriptum.com

Printed at: see last page
ISBN: 978-620-8-18637-1

Antonina Kuzhym

Technology incubator as an ecosystem module

Technology incubator as an organisational solution within the ecosystem of the industrial production development model

Figure 1
Example of a workplace in a startup operating as part of a technology incubator.

Annotation

For the innovative development of society, an organisational solution has been found, which, while continuing its optimisation, modification and improvement, is, judging by the first results, the most encouraging and promising. This solution is the organisation of the so-called technological incubators where new ideas are crystallised and brought to the level of perfection and commercial readiness determined by the market. At the same time, the era of high technology was beginning, which made it possible at the beginning of this process to correctly formulate the main goals and objectives for the organisation of the first innovative structures, which were called, by analogy with greenhouses in which heat-loving plants are grown in greenhouse conditions, technological greenhouses, later renamed technological incubators.

Keywords

Technology incubator; Ecosystem; Effective implementation; Innovative model of industrial and agricultural production development; Structural arrangement of ecosystem base; Quantum computer; Visualisation of augmented reality; Testing laboratories; Laws of technical systems development; Standardisation; Technology transfer.

Table of Contents

Introduction

Laws and practice of integrated development of integrated technical systems and supersystems in conditions of application of quantum computers and their simplified modifications in combination with built-in elements of artificial intelligence and artificial neural networks.

Due to the key factor that determines the efficiency of innovative undertakings and the correctness of the choice of patent and licensing strategy for the development of innovative projects with a high level of world novelty of technical and software solutions used, investors and managers of new (one can say - pioneer) projects face the lack of the necessary level of speed and depth of instant analytical processing of information when modelling processes on modern computers.

Information about the positive results of tests of a quantum computer created at Google Corporation has given reasonable hope for progress in this direction. According to the definition and basic information available, a **quantum** computer is a computing device that uses the phenomena of quantum mechanics to transfer and process data.

(A quantum computer (unlike a conventional computer) operates not with bits (capable of taking the value of either 0 or 1), but with cu-bits having the values of both 0 and 1 at the same time.

Theoretically, this allows all possible states of systems at all levels (both supersystems and subsystems) to be processed simultaneously, achieving significant superiority over conventional computers.

In addition, the lack of such analytical tools currently significantly increases the cost part of the budget of innovation projects, as it requires significant expenditures on computer modelling and high-speed search of options with analytical characteristic assessment of their acceptability and all aspects of efficiency.

Besides, the laws of technical systems development formulated in TRIZ (theory of inventive problem solving) cannot reflect the whole variety of tasks, functions and attributes of a modern multifunctional object, and taking into account all the new and emerging factors that characterise an innovative object, it is necessary to redefine these laws, linking them with the laws of development of commercial structures and commercialisation of innovative ideas.

The formulations should be based on the obtained data and test results of quantum computers, taking into account the nature of the basic contradictions identified in the innovative object .

The dialectical approach (analysis of contradictions) embedded in the main tool of problem solving, which was ARIZ (algorithm for inventive problem solving), was distorted by the introduction of new concepts (technical and physical contradiction) for modelling of which in real time mode the speed and power of existing computers were insufficient.

These new concepts somewhat distorted the essence of the dialectical contradiction formulated in dialectical logic, which led to difficulties in identifying the contradiction when trying to solve real inventive innovation problems with the help of ARIZ, due to the lack of the necessary resource of speed and depth and scope of modelling processes and apparatuses.

We should focus on this separately, but there is an extremely important question of principle - what can be considered a real innovative inventive task?

From the point of view of an investor or project manager, analytical assessment of the project development requires a clear orientation of the situation and analytical modelling according to the following scheme: how a correct or incorrect formulation of the inventive problem may affect the commercialisation of the invention that has arisen?

Correct or erroneous formulation of project tasks and goals, in the absence of step-by-step modelling, may lead to misunderstanding of the classic question - is it possible to reliably protect the resulting technical solution from unauthorised copying?

The search for answers to all these and many other questions is now becoming a major part of the dialectic of creating a development strategy for innovative commercial projects and reliable patenting and licensing of inventions created within the framework of project implementation at all stages and phases of development.

A summary of the structure, activities, research and production linkages of the existing technology incubators is offered below .

Government structures designed to organise, finance and help innovative projects to make their own way in the turbulent commercial sea.

Technology incubators are run by the Ministry of Industry and Trade and partly by the Ministry of Agriculture .

Both of these ministries have a special budget for financing the activities of technology incubators, with a special budget for investing directly in the project and a special budget for financing project management.

Chief Scientist of the Ministry of Industry and Commerce - this position and the related department exist to provide methodological assistance to technology incubators and breeding, and to supervise projects in technology incubators.

The Chief Scientist's Department has deployed units and teams of highly professional experts, both full-time and part-time, and they bear the main load and therefore personal responsibility for the projects they recommend for inclusion in the thematic plan of the technology incubators.

Each of the experts is assigned several projects in different incubators and each expert is a key figure in overseeing and funding these projects.

Export Institute.

The Export Institute provides projects in technological incubators with all the necessary methodological and information assistance. In addition, the Export Institute provides on a competitive basis large grants (about 120 thousand dollars) for the most promising projects for export.

The Institute also assists in the production of prospectuses and other information materials, and promotes newly created innovative products to global markets.

State Standards Institute.

Since compliance with international standards is one of the main conditions for promoting an innovative product on the market, there is a stable practice of continuous co-operation and cooperation between technology incubators and the Standards Institute.

Not least in this respect is the impeccable quality of the evaluation and control of the technical characteristics of innovative products, which is carried out in the laboratories of the Standards Institute using the most modern and efficient control and measuring equipment.

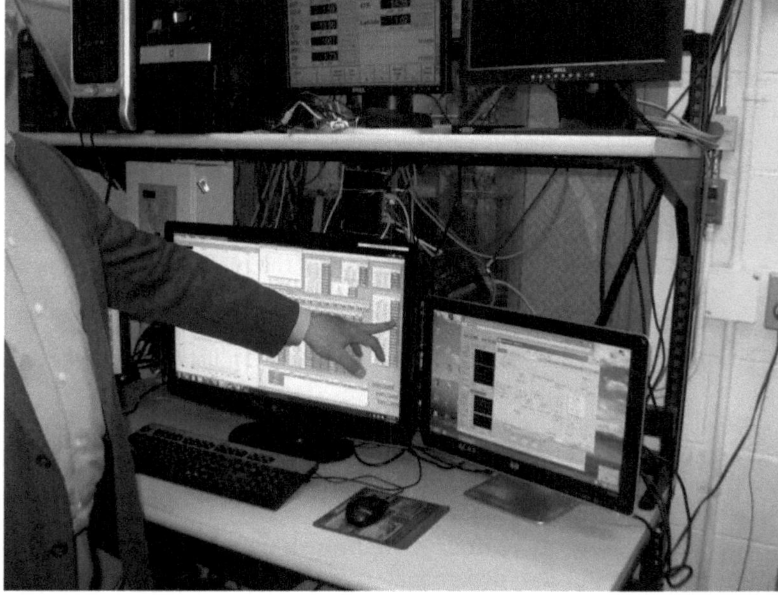

Figure 2

Combined workplace of a startup employee in a technology incubator.

Commercial Testing Laboratories.

In addition to the standards institute, there is also a network of commercial testing laboratories that can test an innovative product according to current standards that are not accepted in Europe, for example.

Also, sometimes standards corresponding to international standards, such as measures of weight or linear dimensions, are not accepted or applied in the countries to which the innovative product is to be exported.

In order to fill this niche and to obtain information about compliance with local standards at the product development stage, technology incubators take full advantage of the benefits of co-operation with commercial testing laboratories

State-owned monopoly companies.

In many countries, national infrastructure facilities are controlled by state-owned monopoly companies.

In order to avoid any friction or misunderstandings, if the innovation project is related to the objects of the national infrastructure and must take into account all available features, technology incubators, in case of technical or commercial necessity, either involve monopoly companies in full partnership or conduct their work on the project in full compliance with the requirements of monopoly companies.

Figure 3

An example of a startup product within a technology incubator. The product is an automatic production line for electronic components, and the same line includes modules for automatic production of parts and a conveyor for assembly and testing. This is an important element of novelty and is the basis for integrative patenting under US patent law.

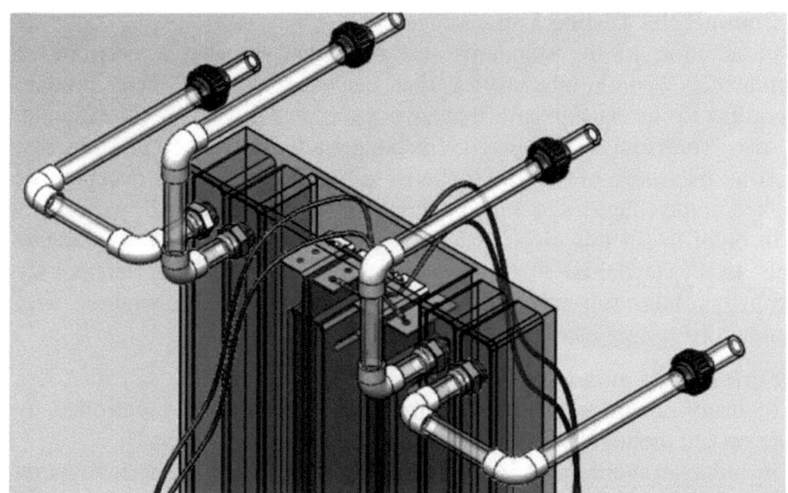

Figure 4

Example of a startup product design as part of a technology incubator. The product is an innovative electrochemical reactor with two parallel electrochemical cells. The treatment process is carried out in an upward flow of the treated liquid or emulsion. This is an important element of novelty and is the basis for integrative patenting under US patent law.

Figure 5

Example of a startup product design as part of a technology incubator. The product is an innovative electrochemical reactor with two parallel

electrochemical cells operating in a fully automatic cycle with a robot leading unloading and loading. The treatment process is carried out in an upward flow of the treated liquid or emulsion. The flow is stepwise, allowing the assembly of the line from a sequential series of working modules. This is an important element of novelty and is the basis for integrative patenting under U.S. patent law.

Figure 6

The figure shows an example of the development and illustration of an innovative product of a startup working in contact with the laboratories of a large technological university.

Interstate and regional investment funds.

Technology incubators take full advantage of every opportunity to finance projects in a timely manner. For this purpose, various regional (European) and US-Israeli specialised investment funds have been set up to finance joint projects and research (such as the BERD fund).

Since each fund has its own funding standard and a set of requirements for determining the eligibility of projects for funding, technology incubators, when forming contracts with new companies in their structure, take into account in advance the possible requirements of these funds and prepare new companies for the time when they may need additional investment.

Municipal Services.

In the localities where technology incubators are located (which are significant employers for small communities), the local municipal services try to assist the technology incubators with any organisational issues that are to a greater or lesser extent within the competence of the municipal services, so that when new innovative companies are formed, the administration of the technology incubator provides for the creation of a logical and material basis for this kind of cooperation.

Public professional associations.

Public professional associations, such as the associations of engineers and architects and many others, try to draw the attention of their members to the problems solved in technology incubators and the professionals belonging to these associations, especially retired people, try, if possible due to the specifics of the project, to pass on their experience in solving similar technical and commercial problems.

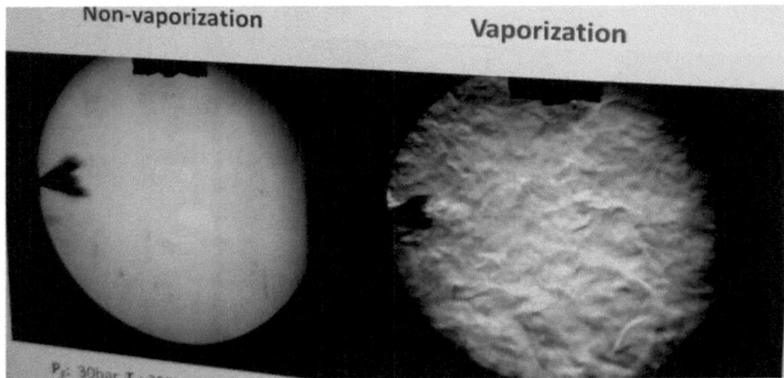

Figure 7

The figure shows an example of the development and illustration of an innovative product of a startup working in contact with the laboratories of a large technological university.

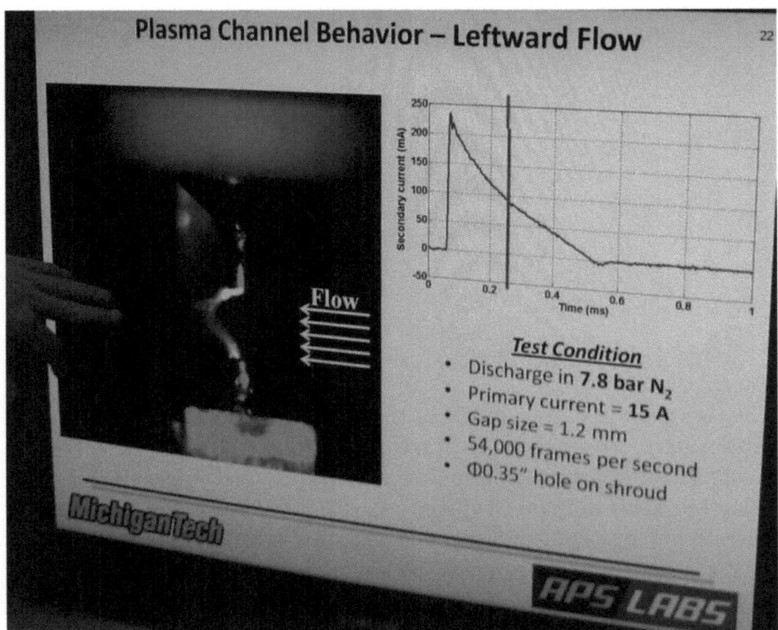

Figure 8

The figure shows an example of the development and illustration of an innovative product of a startup working in contact with the laboratories of a large technological university.

Technical Universities.

Many universities have established their own technology incubators to which promising projects by students and researchers accredited by those universities are primarily transferred.

The conditions in technological incubators are relatively unified and technological incubators in universities only differ in having a younger staff and more comfortable conditions for project development, as the laboratory facilities of universities are also used for the implementation of projects in university technological incubators.

Large multinational corporations

Such corporations with the most concentrated financial, production and technological resources can move innovations forward the fastest, but naturally their interests are purely selfish - they promote only what is interesting, profitable and important to them. Even under such conditions, the role of such corporations and their participation in innovative development and, accordingly, in the development of technological incubators is difficult to overestimate. It makes sense to discuss the specifics of working with such corporations in the innovation field separately in a new publication.

Figure 9

The figure shows an example of the development and illustration of an innovative product of a startup working in contact with the laboratories of a large technological university.

Selecting ideas to form projects.

Each technology incubator today has its own specialisation, although in the early stages of incubators, one could develop projects of several technology areas.

With the emergence of specialisation, and especially from the moment of partial (and in some technology incubators, full) privatisation, the selection of ideas and projects became more systematic, which made it possible to identify projects problematic from the point of view of commercialisation at the stages of their technological expertise. This made it possible to concentrate financial resources on more realistic products from the point of view of market success and ensured significant financial savings.

Existing incentive and assistance systems for the first period of project development.

The rules of work of technological incubators first of all stipulate the schemes of organisation of work of projects and their technological development in the most sparing mode, especially at the first stages of development. This has already been sufficiently discussed in this article, and we can give only one example of how the system of granting an advance of 10 per cent of the total funding for a new project provides an opportunity to attract the necessary specialists to work on the project from the first moments of its development.

International industrial exhibitions and their impact on the successful promotion of projects.

To compare the main commercial, consumer and technical indicators of an innovative product or technology that has become the basis for the formation of a new company within a technology incubator, the best basis for such a comparison are exhibits of all kinds of industrial and commercial exhibitions.

The analysis of such exhibits is one of the tasks facing the commercialisation specialist and, on how he reacts to the results of such analysis, depends the path of development of the commercial qualities of the product, which may lead to a competitive victory, or may lead to failure, in case of unprofessional and inadequate evaluation .

Formation of specialists in technical and commercial expertise of projects.

The specialisation of innovation projects and the need to evaluate them, the conclusions from this evaluation, which are provided to investors and strategic partners in industry and agriculture, have created a market need for commercialisation specialists .

A great many such specialists have been trained in universities, but today's market is driven and dominated, simply by talented individuals who both shape and create commercial success for an innovative product.

Very often they are the ones who synthesise an idea and propose it to the technical specialists, who only have to, as they say, package the idea into a product. According to the experience of recent years, these are the most successful projects in commercial terms.

Technology Transfer.

International projects are often opened and successfully implemented in technological incubators. The scheme of co-operation with technology greenhouses in Israel and other countries, as the authors imagine it to be, is presented to the readers' attention. This does not apply to projects that are already being implemented within a company and have the necessary investments or own funds for this purpose.

Nor does it apply to those projects that are transferred for any innovative company or will be proposed by any innovative company for implementation in any technology greenhouse currently operating in Israel.

The proposal divides the field of any technology hothouse in Israel into work in Israel directly and international collaboration, given the exceptional interest that is being shown today in the market for smart technologies using elements of artificial intelligence and artificial neural networks, the U.S. Federal Government has introduced a programme in 4 areas, - for open research on artificial intelligence; - for research on artificial intelligence with resources that ensure privacy and b

Figure 10

The figure shows an example of the development and illustration of an innovative product of a startup working in contact with the laboratories of a large technological university.

Figure 11

The figure shows an example of the development and illustration of an innovative product of a startup working in contact with the laboratories of a large technological university.

Figure 12.

The figure shows an example of the development and illustration of an innovative product of a startup working in contact with the laboratories of a large technological university.

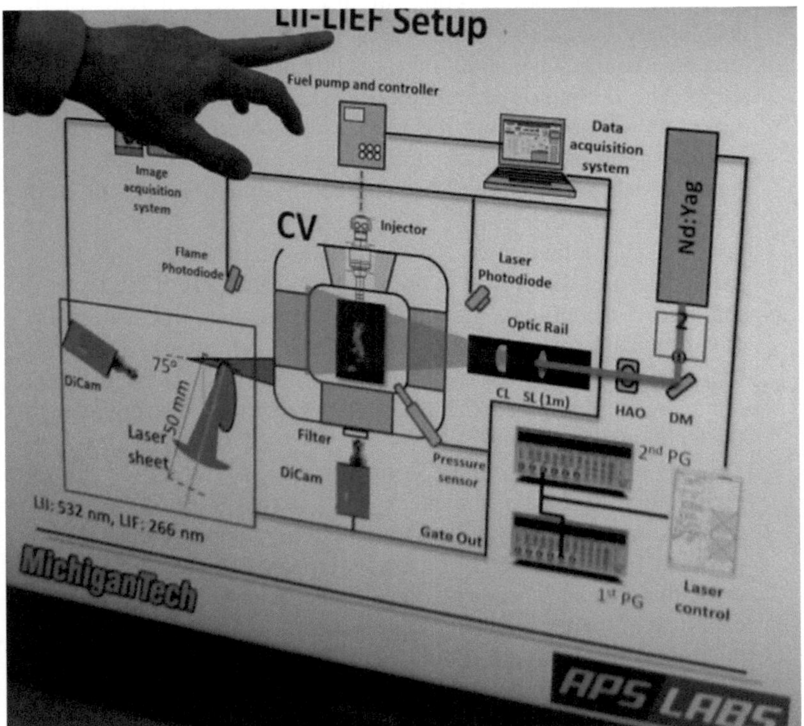

Figure 13.

The figure shows an example of the development and illustration of an innovative product of a startup working in contact with the laboratories of a large technological university.

In the scheme of work in the market of new technologies, the basic idea of which originated, for example, in the USA, the following sequence of organisational and technical stages is proposed:

1. The initiator of the project or the author of the idea finds a primary private or any other investor of at least $50,000.
2. They come to a technology incubator (through some American or Israeli company), where a project is organised under certain standard conditions developed in the technology incubator and agreed and approved by the Chief Scientist of the Israeli Ministry of Industry and Trade.
3. The technology greenhouse (Israel) or technology incubator orders an intermediary company, also on standard terms, to carry out all the necessary work for the initial reality check of the idea and its subsequent patent protection.
4. The intermediary company transfers all materials to the technology incubator; which pays the intermediary company a certain amount of the initial investment amount.
5. The Technology Incubator offers authors and their first investors several options for further development of the project depending on the degree of success of the project at the stage of its analysis and verification in the technology incubator or any other competent structure.
6. If all parties agree, and if the technical level allows it, a presentation of the project is prepared for the Board of Directors of the Israeli technology greenhouse or technology incubator. The Board of Directors decides on the options for further promotion of the project and, if everything meets the basic technical and commercial requirements, decides to open an innovative company in the technology incubator.
7. For the intermediary company, the payment for all the preparatory work and technical expertise of the project is the interest in the company opened in the Israeli technological hothouse and the payment from the budget of the newly opened company for all the work that the intermediary company did to submit the project for expertise in the Israeli technological hothouse, for the translation of the project materials into Hebrew etc.
8. All parties, after deciding to open a new innovative company in principle, must also agree to the subsequent location of the company in Israel's Priority Development Area A or B, in order to receive all the benefits provided for by law (the main provisions on benefits include - full exemption of the innovative company from income tax for a period of 10 years. Refund of about 36% of all costs for the purchase of production and laboratory equipment; assistance in paying the labour costs of production staff during the first year of

operation, etc.). The intermediary company can also work with the Israel Export Institute and seek grants (up to $120,000 per year) for projects with significant export potential.

9. In case the marketing model of a new company geographically gravitates to countries and regions that are not typical for cooperation with Israeli companies, any other scheme of development of the company and its product is accepted.

10. The technology incubator, together with the intermediary company, can in turn test ideas and projects submitted by other Israeli technology greenhouses and implement them in non-traditional regions of Israel on the basis of separate agreements and payment terms.

Stages and stages of project promotion from idea to commercial realisation.

This is a very special issue and it varies significantly depending on the specifics of the project and its place in the market. It can be said that this process in innovation projects does not differ much from the classically accepted forms and is optimised depending on many factors. In practical implementation, the definition of the nature and sequence of stages and phases is very often carried out in accordance with special tables developed by the initiators and about which there are more detailed targeted publications on innovation websites.

Co-operation with the patent attorney community.

Since patent strategy and patent protection of innovative initiatives are an extremely important part of the innovation process, co-operation with patent attorneys and the selection of the right attorney is one of the conditions for the successful promotion of the project, especially in the first stages.

In order to provide this process with a more predictable form and a more confidently predictable end result, a patent and licensing strategy is typically developed before going to patent attorneys, taking into account many of the specific factors inherent in the innovation.

In order to obtain more favourable conditions for this work and to obtain higher indicators in the technological and legal plane, the principle of a high level of competition between lawyers is fully used, which ensures the necessary result.

Collaboration with specialists in commercialisation of projects and technical ideas.

In fact, commercialisation specialists of different schools and levels of competence are constantly present on the innovation playing field, and their correct competitive choice can largely determine more favourable conditions of project financing and a shorter and more successful way of its commercial implementation. The choice of such a specialist is an important process and in many respects the success of such a choice is determined by the preliminary selection of a company that can help in the evaluation of the innovation project itself in its correct technological classification.

Requirements for projects that are developed in technology incubators.

The following documents and materials should be contained, or are recommended to be contained, in the project materials when submitted to the mediator by the company:

Composition and structure of intellectual property objects belonging to the company or initiative group - project applicants. As a rule, all technological areas, which are at various stages of development in the company - project applicant, are complex objects of intellectual property.

Each technology area shall be represented by a systematic structure of constituent elements that include the following principle documents:

1. Forecast of technological development of the direction in the near and distant future.
2. Patent and licensing strategy for all products of the line of business for all stages and phases of the project and for all stages and phases of production and marketing.
3. The principle and basic patents for the inventions underlying the technology direction .
4. Application patents arising from the development of a technological trend.
5. Inventions created by employees and partners of the project applicant company, prior to the organisation of the applicant company, for which there are copyright certificates in the name of the employee or partner, or other legal documents
6. Reports on research and development work carried out by employees and partners of the company - the project applicant in this technological area not within the company.
7. SolidWorks model's on all modifications of products of technological direction, including all variants of assemblies, units, parts and models for digital and virtual simulation of product operation.
8. Basic principles of manufacturing technology of parts and assemblies of products of technological direction.
9. A complete set of working design and process documentation for the manufacture of products of technological direction .
10. Adjusted programmes for CNC machines, designed for the production of prototypes of products of this technological direction.
11. Programmes and methodologies for all necessary types of prototype testing and testing of process direction products at all

stages and phases of production.

12. Actual models of innovative products of technological direction and results of field trials of technological products of agricultural direction.

13. Operational and accompanying documentation for technology direction products, including technological instructions for installation, preservation, storage, repair and transport.

14. Actual prototypes of technology direction products.

15. Technical descriptions of principles and devices of special technological equipment for manufacturing of basic parts and assembly of products of technological direction; materials for registration of patent applications for the specified special technological equipment .

Of course, all these are approximate contents of documents and for each project, depending on its specifics and conditions of implementation and realisation on the market, on the requirements of potential investors, this list may change significantly.

Particularly noteworthy are projects in the field of biotechnology and genetic engineering, in all likelihood a special expert-technological and commercial excursion is needed for this kind of projects.

Figure 14

The figure shows an example of the development and illustration of an innovative product of a startup working in contact with the laboratories of a large technological university.

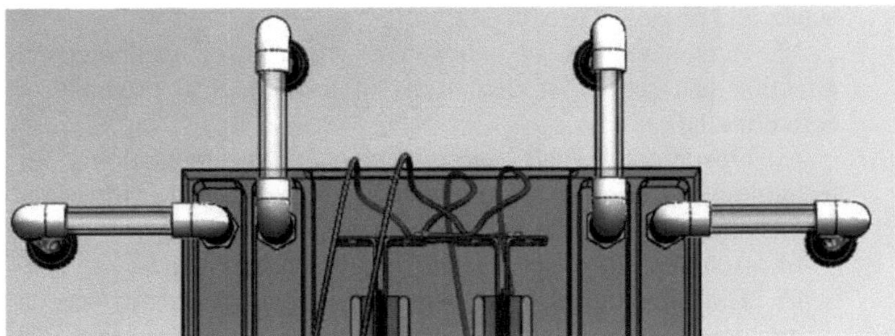

Figure 15

Example of a startup product design as part of a technology incubator. The product is an innovative electrochemical reactor with two parallel electrochemical cells. The treatment process is carried out in an upward flow of the treated liquid or emulsion. This is an important element of novelty and is the basis for integrative patenting under US patent law.

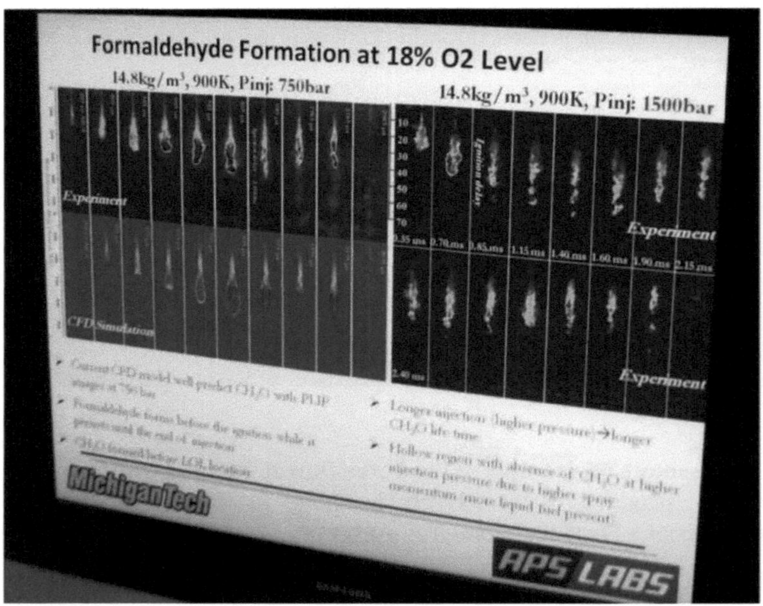

Figure 16

The figure shows an example of the development and illustration of an innovative product of a startup working in contact with the laboratories of a large technological university.

Proposal on options for organising technology incubators in

Ukraine.

Ukraine has all the necessary conditions and prerequisites for the organisation of technological incubators.

Thousands of qualified specialists are working; over the years of stable operation of the enterprises of the military-industrial complex, huge production and technical experience has been accumulated; over the years, engineering personnel have been brought up, the level of professional expertise of which is at the peak of requirements for the initiation of innovative technical solutions.

Ukraine has accumulated a unique experience of agricultural production, which in combination with special, unparalleled in the world, natural conditions (just one example - the famous Ukrainian chernozem) create prerequisites for the creation of agricultural enterprises with the highest level of efficiency and profitability.

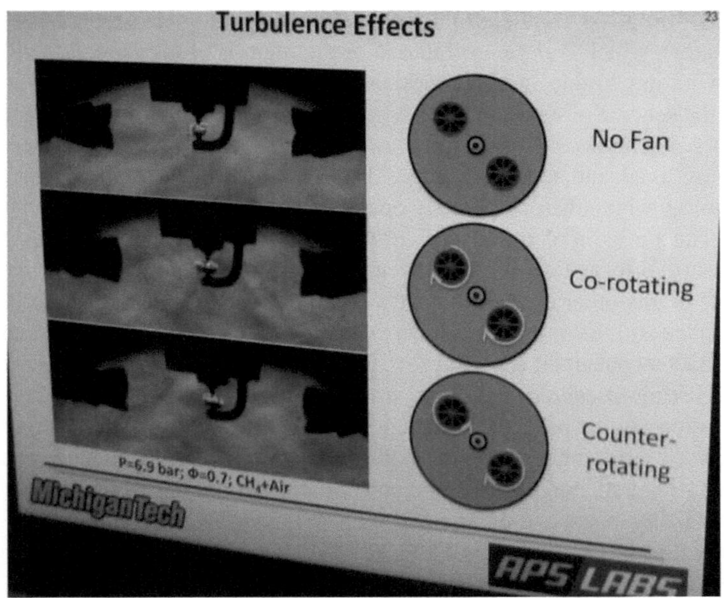

Figure 17

The figure shows an example of the development and illustration of an innovative product of a startup working in contact with the laboratories of a large technological university.

Of course, it is impossible to cover the whole Ukraine with a network of technological incubators at once, and it is not expedient, because with all its technical and commercial revolutionary nature, the creation of highly effective technological incubators is still an evolutionary process. The organisation of the first steps in the formation of technological incubators in Ukraine seems possible and expedient to conduct in the following sequence:

- First of all, based on the opinion of Ukrainian specialists, it is necessary to identify the area of technology in which the total scientific, commercial and historical potential of Ukraine is the highest and most competitive;
- It makes no sense to make any premature and anticipatory forecasts and assumptions, and as it seems to us the most correct, it is necessary to announce competitions for the best technological project in all spheres of production, primarily agricultural, in all regions of Ukraine; to assess the technical and commercial level of the proposed projects, it is advisable to form the terms of the competitions and an expert group, to which it is very useful to invite - for the first stage of evaluation - specialists in commercialisation of innovative solutions from the United States (preferably from the United States).

In the proposal, the field of activity of the technological greenhouse (or incubator), which is organised in one of the regions of Ukraine, is divided into work in Ukraine and international cooperation;

In the scheme of work on the market of new technologies in Ukraine, the authors or initiators of the project are offered the following sequence of organisational and technical stages for its formation (provided that the technological incubator is already open):

1. The initiator of the project or the author of the idea, prepares the necessary information and technical documents and finds the primary private or any other investor for the amount of at least 50 thousand dollars. In case the authors or initiators have difficulties in the search, it makes sense to contact a consultant company.

2. Having received the first private investor, the authors of the innovation, quite possibly together with the investor or the investor's authorised representative, come to the technology incubator, where, under certain standard conditions developed in the technology incubator, the project is organised and on its basis - a new innovative company.

Now it makes sense to briefly discuss these standard (for this technology incubator) conditions.

By analogy with the existing experience, investments in the new company should include private investments (in the amount of approximately 50 thousand dollars, but this is the necessary minimum, and private investments can be even more significant in case the private investor has full confidence in the project). The remaining funds required for the implementation of the project are provided by the technology incubator (this, as already determined from practice, is approximately 400 thousand dollars).

Both the private investor and the incubator do not deposit all the funds at once - first the company receives an advance of 10% of the total investment amount, and as the project progresses, provided that the planned stages are fully completed, the funds corresponding to the cost of each stage are deposited, thus investors insure themselves against any failures or errors of technological nature during the project, because in case of errors and problems, the next financial tranche is made only after the elimination of errors and only with positive results of the stage.

How are the company's shares distributed?

When the company is organised, the authors of the project receive 50% of the shares, the private investor who contributed $50,000 receives 15% of the shares and the remaining shares (35%) are distributed between the technology incubator (25%) and the employees of the company who contributed the most to the project (10%). Refunds are made after the first profit from the project and no more than 5% of the profit per year .

In case of failure and inability to get commercial profit from the project, the funds invested in it are not returned. Naturally, in order to attract investors, state and government structures should find appropriate incentives, the most important of which should be the exemption of a new company from income taxes for at least 10 years after the first profit.

All of the above, of course, are only assumptions and in reality there may be other options for the organisation and implementation of the innovation project.

Now let's assume that the project is successful and cost-effective. In this case, there are additional opportunities for its commercial realisation. This applies to the options of entering foreign markets with the project and, again, if the circumstances are favourable and successful, bringing the company to the stock markets of the USA or European countries.

Who can become a partner in organising technology incubators in Ukraine.

Naturally, the most important and main partner is people, specialists, initiators of innovative ideas and projects. The second necessary and also very important partner and partly organiser are governmental bodies, local self-government bodies, regional and district administrations and possibly legislative bodies.

The third professional partner, are patent specialists, patent attorneys, professional patent attorneys and patent attorneys.

The fourth partner is standardisation bodies and laboratories, control and measurement laboratories, state mining technical supervision bodies, etc.

The fifth partner, are Academic and sectoral research institutes, educational higher and secondary specialised universities and institutes .

The sixth partner, - basic industrial enterprises that have maintained their production and retained in the market.

The seventh partner apparently does not yet exist in reality, but it is a very important partner: investment venture funds and other investment institutions, which should emerge as a result of certain legislative incentives and a system of financial and tax preferences.

While the seventh partner is being organised and formed, there are prerequisites for attracting similar venture capital funds and private independent investors from abroad to the formulated partnership.

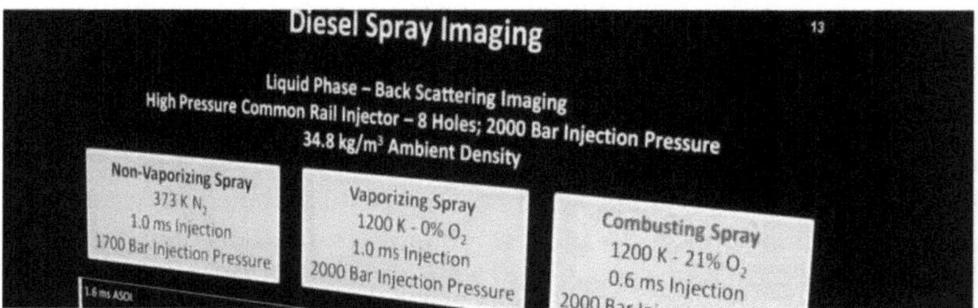

Figure 18

The figure shows an example of the development and illustration of an innovative product of a startup working in contact with the laboratories of a large technological university.

Who it is desirable to invite for cooperation in the organisation and first steps of technology incubators in Ukraine.

As it was announced in February 2011, a visa-free regime between Israel and Ukraine came into force. This means that there will be no need to spend time and money to obtain visas and, if necessary, it will be possible to travel from Ukraine to Israel and vice versa as it is done within Ukraine. Naturally, this will simplify contacts, including at the professional level. Taking into account that every eighth resident of Israel is Russian-speaking and he is a carrier of the same technical culture and mentality as any specialist from Ukraine, it is reasonable to conclude that any professional contacts can be extremely fruitful, productive and useful, will not require additional costs for translators, etc.

If we analyse the experience that Israeli specialists have in many branches of industry and agriculture, in which there is a potential interest in Ukraine, it is possible to single out some of the most important and interesting ones, which in case of cooperation can give the fastest and highest results.

This is experience in the creation of superhard materials and composites based on them, experience in the creation and effective use of electron microscopes, experience in the creation of unique carbon composites and materials and finally a huge, universally recognised experience in all areas of agricultural production and animal husbandry.

Medical technology is a special section of the innovation process and it is worth returning to it and addressing the experience of its implementation in a separate publication .

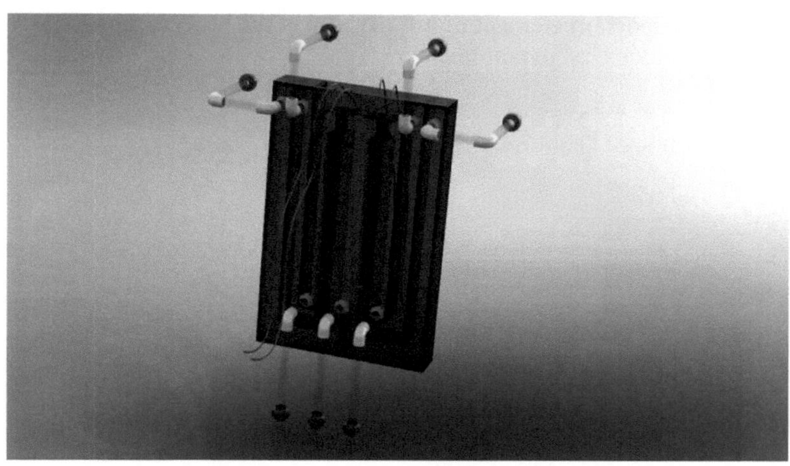

Figure 19
Example of a startup product design as part of a technology incubator. The product is an innovative electrochemical reactor with two parallel electrochemical cells. The treatment process is carried out in an upward flow of the treated liquid or emulsion.

This is an important element of novelty and is the basis for integrative patenting under U.S. patent law.

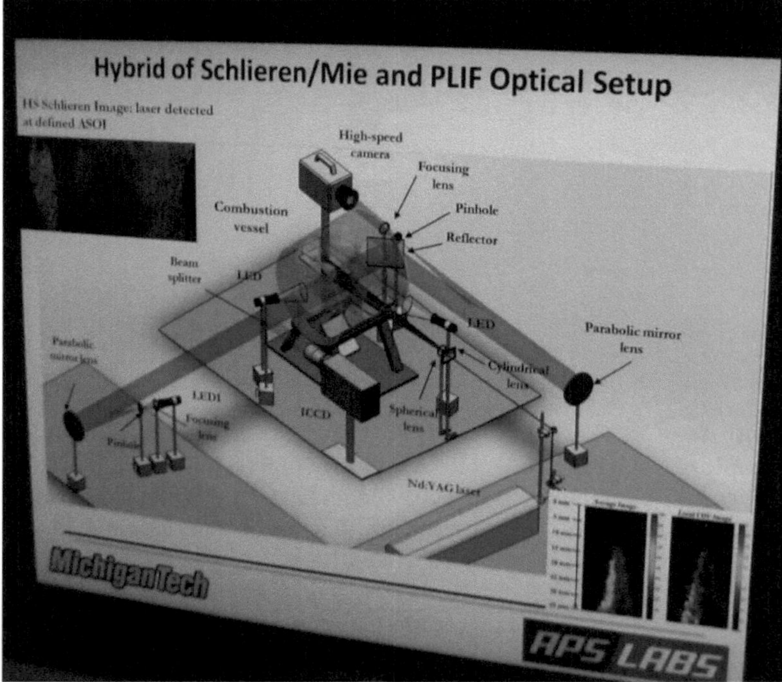

Figure 20

The figure shows an example of the development and illustration of an innovative product of a startup working in contact with the laboratories of a large technological university.

Inventions that open up an innovative service science and technology trend that defines augmented reality visualisation and the possibility of its widespread use.

Augmented Reality technologies in consumer and personal optics are still in their infancy, but given the possibilities for horizontal and vertical integration of these technologies and their derivatives in personal and public infrastructure needs, they are expected to have a great future.

A headset using the aforementioned augmented reality technologies and their primary applications and combinations is still expensive and uncomfortable to wear all day, but the predicted and anticipated upsurge in consumer interest in these groups of technologies and their combinations is forcing an active analytical review of this topic.

Since the service and maintenance aspect for products based on such and similar groups of technologies and accompanying various innovative technical solutions has not been solved, the author of this article, taking into account his experience in the organisation of service maintenance of various household and personal optical applications, considers it possible to make some suggestions and ideas in this direction.

Augmented reality (AR) refers to the technology of superimposing virtual objects on the picture of the real world in the eyes of a person.

The most popular consumer applications of the technology are the DR browser and DR glasses. In the browser, virtual objects are superimposed on a smartphone or tablet camera image, while in the glasses they are displayed on a lens or mini-monitor. Specialists and experts continue to analyse the prospects for possible commercial success when products based on these technologies are introduced to the market, and in the process of discussion interesting details and circumstances are revealed, the correct understanding of which can provide real help in achieving a comprehensive positive result.

Since the release of the first relatively successful project of augmented reality glasses, a lot of time has passed, but there is still no clear idea - what should be the optimal design version of such glasses. These glasses from the first seconds of use caused a growing effect of incomplete understanding of their role and influence on the intensification of the commercial situation on the market of similar products, but despite all expectations, after a certain time, when users had enough time to learn how to rationally use and play with the new gadget, interest in it disappeared and understanding of the situation did not arise.

The fact is that for all this time, developers have not been able to make a device from augmented reality glasses that would be as familiar as a smartphone, game console or smartwatch. As it seems to the author of this publication, one of the reasons for this situation is the state of service technologies, which are not yet ready to service such complex devices in the conditions of mass production and, consequently, consumption.

The catalyst for interest in the technology and an indicator of the relevance of the technology is the current real state of patent protection of these new technologies, which, when analysed in detail, also raises many questions.

What is surprising is the fact that despite the fact that augmented reality visualisation projects and technologies directly and in combination with related technologies can, when implemented and integrated with digital technologies already present on the market, open up new technological directions and promise rapid growth in sales volumes, the dynamics of patent protection is rather sluggish and inactive.

When conducting a patent keyword search, only 133 patent applications were found by the author for the US Patent Office and no patent grants were found on this topic. **Glasses of additional reality, - only 113 patent applications were registered. Visualisation of additional reality, - only**

20 patent applications were registered.

The first analysis of the published patent applications showed that all these applications are of a search nature and do not focus on the design and systems of augmented reality glasses as established products, especially since all examples refer to products at the last stage of development, which indicates a real lack of experience in service and operation

As the author of this series of articles is engaged in the practical implementation of these products on a specific market, it is noted that for such implementation, as of today, a number of technical solutions for products and tools accompanying these innovations are lacking, the first of which include technical solutions and products designed to accompany the augmented reality glasses in everyday life, such as cleaning the glasses from inevitable in everyday use contaminants, both organic and non-organic.

Since the glasses of augmented reality glasses are as dirty as the glasses of ordinary glasses, from a practical point of view, the first thing that should be envisaged is the creation of a complex technology for obtaining a cleaning and disinfecting liquid from water that does not contain high concentrations of salts, including hardness salts.

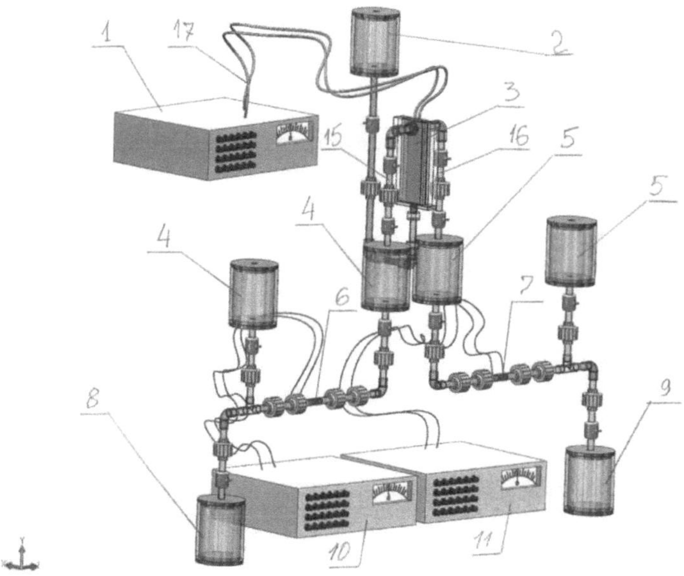

Figure 21

Model system for bi-directional correction of acidity and alkalinity in de ionised water.

1 - power supply for electrochemical reactor

2 - tank with de-ionised water to be fed into the electrochemical reactor

3 - electrochemical reactor with an interelectrode space in which treatment is carried out in two parallel upward flows

4 - collection of water with reduced acidity level

5 - water collector with high alkalinity level

6 - sensor module for measuring reduced acidity levels

7 - sensor module for measurement of elevated alkalinity levels

8 - collection of water with reduced acidity level

9 - water collector with high alkalinity level

10 - pulse generator for electromagnetic resonance sensor

11 - pulse generator for electromagnetic resonance sensor

But this was not enough, as reality demanded **- the** fluid to be used in the washdown must be an insulator to eliminate any possible localised current pulses.

The analysis of possible sources of water for processing in the system has led to the production complexes of photolithography in which de-ionised water is used, having properties close to those properties which we assume to be the most necessary and suitable for the service of optical glasses of augmented reality glasses.

For comparison, distilled water and deep purified water were also considered and analysed, but the properties and qualities of de-ionised water were still preferred.

The possibilities for future use of de-ionised water, both after treatment in the electrochemical reactor and before treatment, in the processes of preparation of various emulsions for subsequent application in service work on augmented reality glasses were also analysed.

Experimental tests have shown the possibility of obtaining high quality emulsions both with de ionised water before treatment and with water after changing the level of neutral acidity or alkalinity .

Figure 22

The figure shows a simplified product model of a product being tested directly in the startup's premises as part of a technology incubator.

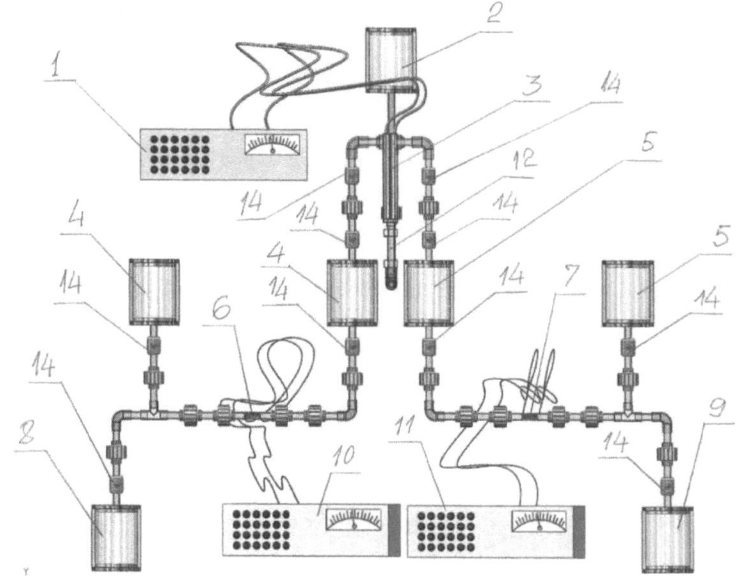

Figure 23

Model system for bi-directional correction of acidity and alkalinity in de ionised water:

12 - inlet pipeline to the electrochemical reactor

14 - control and regulating valves

15, 16 - fluid flow sensors

17 - current supply cables

The selection and analysis of various options finally showed that the most suitable option is de-ionised water, which is widely used in microelectronics and is a familiar component of photolithography technology. This water is produced in significant quantities and its cost is relatively low.

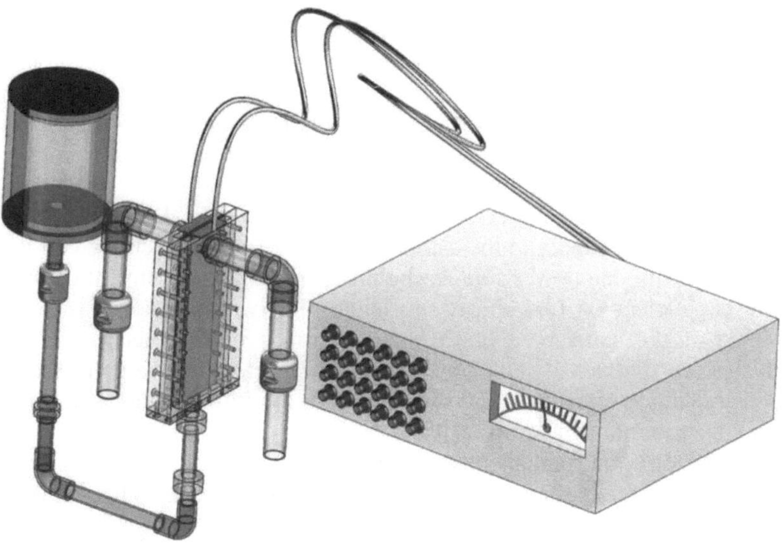

Figure 24

Model of electrochemical reactor with power supply

Since this technology is recommended for use for the first time, it was necessary to produce the simplest prototype to test in real conditions the possibility of practical implementation of several innovative technologies, the first of which was the technique and technology of electrochemical reactor. In the mentioned reactor electrode cells have working zones separated by a neutral membrane located symmetrically between two electrodes. As one of the important principles of operability in the electrode cells, de-ionised water is processed in a developed upward flow. The distance between the working planes of the electrodes is only 3 millimetres, of which the thickness of the membrane is 1 millimetre.

Thus, the thickness of the liquid flow in such an electrode cell is only 1 millimetre. Such thickness of the flow allowed to sharply raise the current density up to 100 amperes per square decimetre, which in turn allowed to perform electrochemical correction of acidity and alkalinity in water very close in its parameters and properties to dielectric liquid.

Since the resulting fluid has two streams at the outlet of the system, it is possible to use both high alkaline water and high acidic water to treat the optics of the augmented reality glasses.

In this way it was possible to clean the surface of optical glasses from microorganisms and bacteria using water with an acidic background and to clean grease contaminants using water with an alkaline background.

It should be noted that since the system for preparation and processing of de-ionised water is very simple in design, it uses the most widely used in mechanical engineering construction materials and components, it can be installed in any small trading company selling optical devices and glasses, including augmented reality glasses, both current modifications and future modifications.

There is one more factor that is not related only to the specifics and requirements for the optics of augmented reality goggles, but in general to optical devices for general use

This is an anti-allergic effect. Since in the electrochemical reactor of the system the current supply to the cathode and anode of the electrode cell is carried out from one power source, the phenomenon of adjustment of the liquid parameters in two directions is absolutely proportional in the presence of the same area of the active surface of the electrodes, at equivalent current density - that is, both acidic reaction of one part of the flow and alkaline reaction of the second part of the flow do not cause any allergic reactions at simultaneous use.

This phenomenon fundamentally changes the consumer qualities of the system not only for augmented reality glasses, but also for any other optical devices .

As in every system for the dynamic electrochemical treatment of liquids, and especially for the treatment of liquids with low current conductivity, the design of the electrode cell is of critical importance. All design features of such a cell, all electrode materials, the materials and design of the neutral membrane and the materials and design of the cell housing are taken into account. As it turned out, a special efficiency potential is inherent in the application of electrodes made of carbon-carbon composites (the phenomenon of application of these materials will be described in subsequent publications).

Figure 25

Model of electrode cell of electrochemical reactor

The simplicity of the electrode cell determined the simplicity of the entire water treatment system.

Figure 26

Photograph of a system for real acidity and alkalinity correction in de ionised water.

Photos of the real system - prototype demonstrate the designed extreme simplicity and as a consequence low cost with very convenient operation. Such a system can be installed and operated in almost any commercial complex without special expenses and with the help of products produced with its help, to bring the level of service of this type of products to the level of technological perfection of all current and future optical devices.

It should be said that, if we make a deeper structural analysis of all the possibilities and characteristics of the mentioned system, we can see that this system, in complex and in many private local solutions, allows us to significantly increase the range of influence on the results and quality of operation of innovative products in modern conditions.

Moreover, the particular case of the above system with minimal modification and optimisation allows to open completely new possibilities, which, in case of system three-dimensional design, can give an impetus to the synthesis of further ideas on system improvement and on modernisation of the system itself in order to obtain a new system opening more promising scientific and technological directions. As an example, let us take the already mentioned scheme of the structure of volume-porous electrodes. Let's take several elements from this scheme and try to predict possible variants for modifications with the final result of obtaining a new product. Let's return to the three-dimensional model of electrodes.

As we have indicated previously one component of the electrodes are parts made of carbon-carbon composite materials. If we consider one version of the electrode design as a composite electrode, when the electrode body has at least two elements, for example, a strip of titanium coated with a conductive polymer coated on the water contact side with a carbon-carbon fabric obtained by pyrolysis and saturation of viscose fabric with carbon.

Such a fabric has excellent conductivity with very good permeability

Products made of such fabric can operate at very high temperatures, but the fabrics themselves can withstand temperatures up to several thousand degrees .

As practice has shown, from such fabric, with certain electrochemical treatment can be obtained napkins necessary in medicine and according to the results of tests showed high results, especially in the treatment and prevention of burns.

Now let's get back to the service - **in** addition to water with acid and alkaline background for cleaning or washing of the lens surface, wipes are also needed. The final processing of wipes can be carried out on the same equipment and the use of such wipes with absolutely neutral effect on the surface of glass and possible variants of polymeric materials can bring an additional impulse of quality to the service of innovative types of glasses and not only.

There are ideas on use of carbon-carbon fabric in many special modules for purification and regeneration of water and aqueous solutions that forms a prospect for more complex introduction of these technologies in the technique of ion-exchange water purification with application of natural zeolite as an ion-exchange material.

If we continue modelling the situation, we can propose the construction of capsules made of carbon-carbon fabric with a load of zeolite granules and the capsule (carbon-carbon fabric is an excellent conductor of electricity) can be connected to a power source, i.e. ion exchange purification can be combined in such a construction either with treatment and activation in an electromagnetic field or with treatment under heating conditions.

The search for a free niche in the above-mentioned technological field continues with exceptional intensity, but again the level of this intensity is not completely clear, as the intensity of patent protection of these solutions does not meet the declared level of development of these projects and also the declared level of breadth of search for new technical solutions

A detailed analysis of the patent and licence situation shows that in this group of products and technologies it is very difficult to identify a basic technical solution, as individual technical solutions overlap with each other, and it is practically impossible to determine which of the technical solutions ensures the achievement of an ideal end result.

Besides, in view of the fact that many distinctive features of technical solutions on which the augmented reality technologies are based for many people are quite obvious, but at first sight not necessary, the necessity of which is a figment of imagination, and not an urgent necessity.

In any case, commercial success in bringing innovative products to the market depends directly on the degree of preparation of the entire infrastructure and the consumer to perceive the novelty of this product and the correct formulation of requirements for the new product and also on the correct formulation of the degree of user expectations from the introduction and use of the new product.

It should be noted that, unfortunately, as of today there is no regulatory documentation, including standards and instructions regulating both the technique of operation of such products **and** the procedure of operations to ensure the level of necessary quality of operation of these products and, most importantly**, the** necessary level of safety when using them.

The impact of complex products such as augmented reality glasses on all aspects of user health has not been studied in any way, especially for the long term.

There are serious and well-founded concerns that significant changes in the technical capabilities of the new glasses may lead, among other things, to mental psychological deviations in the perception of reality.

Based on extensive experience in the selection, development and implementation of standard-type glasses, the author of this article believes that the developed user stereotype of application in everyday practice of optical devices, including glasses, will be in consumer dissonance or even conflict with those unusual and specific opportunities that bring innovative optical products of this type, including augmented reality glasses.

Of course, an opponent might object and recommend that users be prepared to properly perceive the new product, to properly use the new product, and to properly support the correctness and performance of the elements of an unusual new product, especially optical lenses.

The author of this article believes that first of all the developers of the mentioned innovative products should start checking all aspects of the market situation from the moment of origin of the idea of a new product or technology and as the degree of success in the project becomes clearer, it is the developers who should be closely connected with insurance companies to determine the degree of danger of new products to the health of users. In principle, the process of such harmonisation between the product developer and its potential users has long been known.

Taking into account the need to maintain a certain level of confidentiality for the authors of the idea and for the developers of techniques and technology, naturally the process of such coordination can be initiated by the fact of preparation of initial technical requirements for a new product and a set of technologies for its use and operation. Of course, all any significant parameters of the initial technical requirements at the stage of their formulation will have to be harmonised with similar or somehow related parameters of the current standards. In this case, the harmonisation can be carried out at the level of the standards institute, which will preserve confidentiality and protect the idea from the curiosity of potential competitors. Thus, in this variant of the organisation of the approval process, the first to answer potential questions will be the standards institute, the qualification of its specialists does not raise any doubts. Nevertheless, the very idea of creating such glasses has serious potential and will be much more in demand if service technologies and basic equipment for their implementation are created in parallel.

Figure 27

The figure shows a test bench in a large company, where the product of a startup working as part of a technology incubator is being tested by prior arrangement.

Figure 28

The figure shows a test bench in a large company where a product of a startup working as part of a technology incubator is being tested by prior arrangement.

Figure 29

The figure shows a test bench in a large company, where the product of a startup working as part of a technology incubator is being tested by prior arrangement.

Figure 30

The figure shows a test bench in a large company where a product of a startup working as part of a technology incubator is being tested by prior arrangement.

Figure 31

The figure shows a test bench in a large company where a product of a startup working as part of a technology incubator is being tested by prior arrangement.

List of used literature and patent and licence information

Annex 1

United States Patent Application	**20180031836**
Kind Code	**A1**
Johnson; Lonny Eric ; et al.	**February 1, 2018**

SMART GLASSES HAVING INTERFERING LIGHT FILTERING

Abstract

A pair of smart ***glasses*** having interfering light filtering includes a ***glasses*** frame defining a horizontal view path and a semi-transparent display supported on the ***glasses*** frame and positioned in the horizontal view path. A projection mechanism is supported on the ***glasses*** frame and has a projection lens positioned above the horizontal view path. The projection mechanism is configured for projecting virtual content on the semi-transparent display, with an outer surface of the projection lens facing a rear facing surface of the semi-transparent display. A semi-transparent polarised shield is supported on the ***glasses*** frame and is positioned below the horizontal viewing path, with an inner surface of the semi-transparent polarized shield facing the outer surface of the projection lens. The semi-transparent polarised shield is positioned to filter interfering light passing through the semi-transparent polarized shield and towards the projection lens.

Annex 2

United States Patent Application	**20170307787**
Kind Code	**A1**
Kawamura; Takumi	**October 26, 2017**

HEAD MOUNTED APPARATUS AND GRIPPING APPARATUS

Abstract

A head mounted apparatus that prevents the influence of external light and is easy to use even for a user wearing ***glasses***. The head mounted apparatus includes a display unit that displays an image to the user, and a light shielding member that shields a periphery of the user's eyes from external light when the head mounted apparatus is mounted on the user's head, wherein an opening is formed on the light shielding member at the user's orbital regions.

Annex 3

United States Patent Application **20150378155**
Kind Code **A1**
KUEHNE; Marcus ; et al. **December 31, 2015**

METHOD FOR OPERATING VIRTUAL ***REALITY*** GLASSES AND SYSTEM WITH VIRTUAL ***REALITY*** GLASSES

Abstract

A method for operating virtual ***reality glasses*** involves displaying at least one virtual object by the virtual ***reality glasses*** from a virtual viewing position and continuously detecting a position of the virtual ***reality glasses*** and adjusting a virtual spacing between the virtual viewing position and the virtual object. Once the virtual viewing position passes through a surface bounding an element of the object from the outside, the representation of the element is altered. Furthermore, a system with virtual ***reality glasses*** may be used.

Annex 4

United States Patent Application **20160034042**
Kind Code **A1**
JOO; Ga-hyun **February 4, 2016**

WEARABLE GLASSES AND METHOD OF PROVIDING CONTENT USING THE SAME

Abstract

A wearable ***glasses*** is provided. The wearable ***glasses*** includes a sensing circuit, a communication interface, a display, and a controller. The sensing circuit senses movement information of a user wearing the wearable ***glasses***. The communication interface receives notification message information. The display displays the notification message information within an angle of view of the user wearing the wearable ***glasses***. The controller determines a movement state of the user based on the sensed movement information of the user and controls the display to display the received notification message information according to the movement state of the user.

Annex 5

United States Patent Application **20180005421**
Kind Code **A1**
PARK; Jisoo ; et al. **January 4, 2018**

GLASSES-TYPE MOBILE TERMINAL AND METHOD OF OPERATING THE SAME

Abstract

A *glasses-type* mobile terminal includes a display configured to display a virtual ***reality*** image and a controller configured to acquire ***reality*** information from a mobile terminal connected to the ***glasses-type*** mobile terminal and controlling the virtual ***reality*** image if a ***reality*** returning time indicating that viewing of the virtual ***reality*** image should be finished is reached based on the acquired ***reality*** information.

Annex 6

United States Patent Application **20170366805**
Kind Code **A1**
SEVOSTIANOV; PETR_._ __ VYACHESLAVOVICH **December 21, 2017**

METHOD AND SYSTEM FOR DISPLAYING THREE-DIMENSIONAL OBJECTS

Abstract

A system for displaying three-dimensional objects using twodimensional ***visualisation*** means simultaneously providing at least effects of binocular parallax and motion parallax, the system comprising: a display configured to display a sequence of images; a pair of ***glasses*** configured to provide stereoscopic separation of images, the ***glasses*** comprising at least two optical shutters and at least two markers; two optical sensor arrays; two reading and processing devices configured to read data from an area of the optical sensor array and to determine 2D coordinates of the markers; a marker coordinates prediction device configured to extrapolate coordinates of the markers so as effective overall delay does not exceed 5 ms; a marker 3D coordinates calculation device; a 3D scene formation device; and at least one image output device. The invention also includes a corresponding method of displaying three-dimensional objects and provides realistic representation of three-dimensional objects for one or more viewers.

Annex 7

United States Patent Application	**20170371164**
Kind Code	**A1**
Liao; Chunyuan	**December 28, 2017**

WEARABLE SMART GLASSES

Abstract

The smart ***glasses*** comprise a ***glasses*** frame. The smart ***glasses*** further comprise a pair of ***glasses*** legs connected to two lateral portions of the ***glasses*** frame respectively, each of the ***glasses*** legs having a first end and a second end, wherein the second end of each of the ***glasses*** legs bends inwardly to form a first arc portion. In addition, the smart ***glasses*** further comprise a pair of clamping members each disposed at one of the ***glasses*** legs, wherein each of the clamping members comprises an elastic clamping element, said elastic clamping element being disposed at the inner side of said second end, wherein said elastic clamping element having a free end that bends inwardly to form a second arc portion.

Annex 8

United States Patent Application	**20160173865**
Kind Code	**A1**
PARK; Sung Woo	**June 16, 2016**

WEARABLE GLASSES, CONTROL METHOD METHODOF, AND VEHICLE CONTROL SYSTEM

Abstract

Wearable ***glasses*** include a first capturer for photographing a front part, a second capturer for tracking a gaze direction of a user, and a controller for matching a target image captured by the first capturer with a three-dimensional (3D) interior map of a vehicle corresponding to the target image, and determining a head direction of the user. The controller specifies an object corresponding to user gaze in the target image based on the determined head direction and the tracked gaze direction of the user. Accordingly, user convenience is enhanced.

Printed by Books on Demand GmbH, Norderstedt / Germany